Not Just Birds!

Seed Spreaders

Written by Diane Kann

Illustrated by JB Barbara

Kannceptual Creations

Paws to Read – A moment shared is a memory made.

Visit us at: ***boltandcharliebooks.com***

Published by **Kannceptual Creations LLC**

An imprint of Bolt and Charlie Books

boltandcharliebooks.com

ISBN: 978-1-972343-48-7

Printed in the United States of America

First Edition, April 2026

Paws to Read is a Bolt and Charlie Books initiative promoting early literacy and shared storytime experiences.

DEDICATION

To all the little readers,
with bright and curious eyes,
who love the trees, the animals,
and every wild surprise.

May you see the helpers
in forests big and small,
from birds that soar so high
to ants that crawl so small.

May you plant tiny seeds
with care in all you do,
and grow a greener, kinder world
that blooms because of you.

With love, always—
from Granda Diane

Do you know who plants the trees?
Who grows the forests, breeze to breeze?

You might say, Birds!—and that's true too...
they carry seeds the whole day through!

But wait! There's more—look close, you'll see... more helpers grow each mighty tree!

A monkey munches fruit so sweet,
juice on its chin, seeds at its feet.

It swings away from where it fed,
and drops those seeds in places new instead!

A bat flies out as daylight fades,
through jungle paths and leafy shades.

It nibbles fruit in the starry night,
then spreads new seeds in silent flight!

**A squirrel gathers nuts with care,
hiding them here... and over there!**

But some are lost (oh yes, it's true!)
and grow into trees both strong and new!

A deer walks softly through the land,
seeds travel with it—just as planned.

From place to place, they journey far, helping forests grow where they are!

Even ants—so tiny, so small—
carry seeds along the forest floor's hall!

They tuck them safely underground,
where brand-new plants can soon be found!

So many helpers, near and far, growing forests as they are!

Not just birds (though they do their part!),
but a whole wild team with nature's heart!

You can help forests—yes, you can!
Plant a seed and lend a hand!

THE END...

Wow, what an amazing story we just shared!

We learned that it's not just birds who help grow forests—so many animals like monkeys, bats, squirrels, deer, and even tiny ants all work together to spread seeds and bring new plants to life.

Isn't it wonderful how nature has its own team of helpers?

Each one, big or small, plays an important role—just like each one of you!

And do you remember the most exciting part? *You* can be a helper too!

By planting seeds, taking care of plants, and protecting nature, you can help forests grow strong and healthy.

So next time you see a tree or a plant, remember—it might have been planted by one of these amazing helpers... or maybe, one day, by you!

Thank you for listening, everyone!

Now, who's ready to be a seed spreader hero?

LEARN THE KINDS OF BIRDS

Macaw

Hummingbird

Owl

Duck

American Robin

LEARN THE KINDS OF BIRDS

Kingfisher

Flamingo

Toucan

Puffin

Meadowlark

www.ingramcontent.com/pod-product-compliance
Lightning Source LLC
Chambersburg PA
CBHW041323210926
51847CB00002B/33
9781972343487